The Compleat Collier: or the whole Art of Working Coal Mines in the Northern Parts;

m.dcc.viij.

Reprinted at Newcastle,
by M. A. Richardson,
in Grey Street,
mdcccrlv.

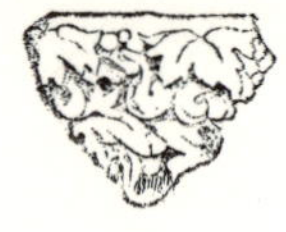

First Edition 1708
Second Edition 1845
This Edition 1968
Reprinted 1979

Published by
FRANK GRAHAM
6 QUEEN'S TERRACE
NEWCASTLE UPON TYNE
NE2 2PL

900409 51 7

Printed by J. & P. Bealls Ltd.,
Gallowgate, Newcastle upon Tyne, NE1 4SA

Advertisement.

F the author of this valuable tract we are unable to recount any thing whatever: from the intimate knowledge he displays of the locality and its peculiarities, we may safely believe him to have been a native. Our author's fears as to the safety of introducing engines for " drawing water by fire," or as we would say, steam, from the danger of bringing " more fire into the deep Bowels of the Earth," where there is already too " much Sulpherous Matter;" and at the same time his wish to have the advantage of their superior powers, were in the first case dissipated, and in the second exemplified about six years after the publication of his opinion,

vi.

for we find that the first steam engine in the North
was erected by the son of a Swedish noble at Heaton in
mdccxiv. In other respects the tract is essen-
tially curious as exemplifying the mode
of working the coal mines at a
comparitively early
period.

Quothe M. A. R. at his dwel-
ling in Newcastle,
yᵉ December,
mdccclv.

T H E

Compleat Collier :

Or, The whole

A R T

OF

Sinking, Getting, and Working, Coal-Mines, &c.

AS

Is now used in the Northern Parts,

Especially about

SUNDERLAND

AND

NEW-CASTLE.

By *J. C.*

Et thus & vividum Ignem in Carbonibus Prudent.

London : Printed for *G. Conyers* at the Ring in *Little-Brittain.* 1708.

T H E

Compleat Collieries, &c.

Ollieries, or the Coal Trade being of so great Advantage to the Crown and Kingdom, I have thought fit to publish this short Treatise thereof, (having not met with Books of that Nature hitherto,) in order to encourage Gentlemen, (or such) who have Estates or Lands wherein Coal Mines are wrought or may be won, to carry on so useful and beneficial an Imployment, as this is, which I need not mention the Particulars of, both in respect of Fires, for private and publick use, as also in respect of the Revenue it brings in
Yearly, or in Respect of the great *And to build* Advantage it is, being a Nursery *St.* Paul's. for Saylers, &c. or lastly, in respect of its imploying so many Thousands of the poor in those Northern parts of *England,* which

are maintained by it, who must otherwise of Course be Beggars, or would be starved, if Coal Mines were not carried on; as also explain the whole Art of Collery, that Coal-Owners, (or such who have or may work their Colleries) may of themselves by this Book know better how to carry on such an Undertaking, and that they may not be too much imposed on by their head Servants, or such who imploy the poor Labourers or Miners in their Colleries; by keeping the Owners in Ignorance of the Work and Charge; therefore in order to proceed on these several Heads, I shall begin (and with as much Brevity as I can) after this plain following Method.

Coal-Owner.]   Friend, you tell me you have hopes to win a Colliery in my Grounds, pray tell me your Reasons for it?

Servant, or Undertaker.]   Sir, my Reasons are as follow: In the first place your Ground Borders on other Colleries, which are working Colleries, which makes it plain that there is Coal so near you; but more especially, you may please to observe in your Grounds, the Undoubted Tokens or Signs of a Collery, which are these following: First, there is an Out-burst or an appearance above Ground, of some Vein of

Signs of a Colliery. Camb. Brittain.

Coal, which some History Writers says, was the first Encouragement to begin Coal-Work: Or, Secondly, you have an Out-burst or appearance of such Stone (as we call Coal-Stone) but if these Signs do not satisfy you throwly: We have in the last place this undeniable Proof and Assurance by *boreing the Grounds* with proper Instruments whereby we can discover the Nature of the Earth, Minerals, and Water, that may be met with in our way of Sinking, nay we can thereby discover to a small Matter how deep your Coal lies in the Earth, and what thickness the Coal-bed is of.

Boreing an Assurance of Coals.

Owner.] Pray Name me your boreing Instruments, and let me know their use.

The way of Boreing.

Servant.] We have two Labourers at a time, at the handle of the bore Rod, and they chop, or pounce with their Hands up and down to cut the Stone or Mineral, going round, which of course grinds either of them small, so that finding your Rod to have cut down four or six Inches, they lift up the Rod, either all at once, as there is conveniency for its Lift; or by Joynts, fixing the Key, which is to keep the Rod from droping down into the hole, and so sometimes looses the Rod, or at least, occasions a great loss of time to get it up again, and taking off the cutting Chissel, puts, or screws on the Wimble or Scoop which takes up the cut Stuff be it what happens; and so by sight of the Stuff taken out of the Wimble, or Scoop, you plainly discover of what Kind it is, which we so cut or grind, and so consequently what follows to any depth; for by Addition of Joynts, (which we screw into the Rod, we can descend to any depth; as also by knowing the length of the Rod to such and such sort of Stone and Coal, &c. we have the certainty of what depth the Coal lies, and likewise thereby may know to an Inch the thickness of the Bed, or Vein, of Coal, besides, (which I need not mention) we also find whether the Rod goes down dry, *Boreing* or where we prick'd the Water Fee- *useful.* ders, and (what is farther useful and requisite in Collyries) we can hereby find whether we are sinking upon an old wast (or wrought Collery) which saves a great deal of Moneys in sinking, so deep as 30, 40 or 60 Fathom to no purpose, for you will not be at above 15 or 20*s. per* Fathom Boreing, when perhaps it may cost 50*s.* or 3*l. per* Fathom to sink at least, nay, many times a great deal more, especially if a Whin (which is the hardest sort of Stone, met with in the Earth) lye in the way to be cut through, which I have known so hard, that the hardest temper'd Steel we could prepare, would scarce work upon, and that the said Whin of two Yards Diameter, and little more than 23 Inches thick has stood in near 20*l*. cost. But, Note, that about 3 Inches Diameter for a Bore-hole (or Boreing) is sufficient.

Coal-Owner.]   Well, I think you have pretty well satisfy'd me of the Method, Benefit, and Charge of Boreing, but do you not use Boreing sometimes in (or with) Sinking?

Servant.]   Yes, Sir, very often, for if we meet with a great Feeder of Water in Sinking a Pit in a working Collery, to save the Trouble and Charge of Sinking with that Feeder to the Coal, for it were Folly and unreasonable Charge, besides a vast loss of time to, to Lave, or fill 20 or 30 Tubs of Water *per* Hour, of 30 Gallons a piece, which Water by a *Bore-hole* may be let down into the Collery, where we design to hole our Pit, and so is by a Drift or Watercourse from the old Pits, set away to the place where your Collery Water is all drawn, therefore such Feder so running down such bore-hole, saves its Charge of Boreing in Sinking, by preventing the cost and loss of time in Laving, the Water or Feeder, that would otherwise be drawn up as they work or sink the Shaft, as is the dayly Practice of these Northern

Parts. And lastly, it is a very *good Caution*, even in a Coaled Pit, to put down a Bore-Rod about a Fathom, for a *Sump* (or Well to a Coal-Pit) to hold the Drawings (or filings as we call them there) of Water, whether Rain or otherwise, least in that little depth, after a Pit is Coaled, we should raise a *Feeder* which would be a great Charge on the Collery to draw, nay sometimes has drowned a Collery by Sinking it rashly, which might have been prevented by a Fathom boreing in the Thill or Bottom under the Coal you would Work; and you may have Boreing undertaken by the great, or otherwise our Sinking Labourers, with a careful Observant Over Man's Advice and Inspection, will perform it for sinking Wages, which is about 12*d.* or 14*d. per* Day.

Coal Owners Object.] But I have heard of Colleries which have been drowned by Boreing, what say you to that?

Servants Answer.] That sure has been for want of Care in the Over-Man Imployed, for we should always be watchful, and have good strong wooden Plugs ready made, whilst boreing, to chop into the Bore-hole immediately if a strong Feeder shou'd chance to be raised of a suddain, which Pluging will stop any Bore-hole Feeder I dare affirm; if that Vigilency be observed duly.

Coal Owner.] Well, let me hear now what you

say to sinking a Coal-Pit, what Utensils are requisite? and the Method you propose in Sinking.

Servant.] First, to Sink a Pit, we must have a stock of Timber prepared, that is to say, we must have either Oaken Spars, or Firr bawks, which is most used (being cheapest in these parts) for Pit Bars, and a quantity of deal Boards to be wedged in between the Bars and the Earth, to keep the Earth, or some times soft Mettle, or Minerals (as we meet with it), from falling into the Pit, and filling up our sinking, with a good Quantity of Nails, likewise the Pick-Ax and Shovel to break up the Soil or Surface of the Earth, we are to dig till we sink down to the *Stone head*, so that the said Pit-Bars of Wood and Deals must be. used till we get to the Stone, and in some places it may be 8 or 10 Fathom of Earth to the *Stone Head* and more, and in other places but a small matter, so that the Provision for such things is proportioned to the depth as needful.

Observations. And now I shall make bold to recite to you something, which in my Experience of Sinking I have Observed, that after we get down to the Stone Head, there being Mould, Sand, Gravil or Clay (all which I call Earth) till we come to the Stone, we never meet with any more Earth again, let us descend to what depth we can, but find it all some sort of Stone (or Rock) or Menerals, and Vains (or Beds) of several sorts of Coal, for you must Note, we have several Coals in some Colleries before we get the Main-Coal, which is generally most esteemed.

Names and sorts of Coals. There is in some Colleries 7 or 8 sorts of Coal. There are the Pipe-Coal, which perhaps is not above 4 or six Inches thick. The Crow-Coal about a Foot or less thick. The 3 Quarter Coal about 3 Quartes thick or more, all which are foul or bad Coals, and not worth much, tho' I have known the 3 Quarter Coals used in making Salt, and believe it is so yet at *Shields* or elsewhere, but, I think the best Coals are best for the Salt Pans and Salt too, and make most and best Salt, if they could afford a Price, as no doubt they may, for best Coals, as some Salt-makers have confest. And there is the half Yard Cole, which is so called, it being of about that thickness, but is generally good Coal, and better than the 3 Quarter Coal, yet being so low to work in, or but of that small Thickness) it is scarce worth while to work it, there is also the five Quarter Coal, which is of about that thickness of 5 Quarters, and that is in some Colleries very fine, and makes a hot quick Fire, but doth not Cement or Cake well, yet I have known it make excellent Fires for any sort of use, but not so lasting as the Main-Coal is in general :

I am told there is at a place called *Cullercoats*, or *Blyth Nook*, a Coal much of the Nature of this 5 Quarter Coal, which I have above described, and have heard that these 5 Quarter Coal, mix'd with Main Coal, give Satisfaction in their Use or Burning : But there is also the Main-Coal, which is in some Places thicker then others, sometimes you have it 7 Quarters, sometimes 8 Quarters, and sometimes 8 Quarters and half thick, any of which indeed is thick enough and valluable ; this is the Coal I would wish you to have, this is the thing I think may please all People for Use, for this in these Northern Parts is generally best, both for a good Fire, and a lasting Fire. You may make a Fire of this 2 or 3 times over, for turn it over after it is Caked, it will again burn brisk and fresh as at first, but its Excellency is so generally known, that I need add no more on this Head, only by the By I think I may venture to say, It is not so much valued as it ought to be at this time, especially at the Pits, for was it ever heard of, or known that this *Noble*, this *Main-Coale*, was sold, as lately it was, or now is, for 8*s. per* Chaldron, Water, or *New-Castle* or *Sunderland* Measure, which (you must Note) is generally reckoned double the Measure of a *London* Chaldron, or more? This I think is shameful for Owners, who striving to get all the Trade to themselves, or to have a *Major* Part of

Vend, will fall out among themselves, and as it were almost give or throw away so valuable and Indispencibly necessary a Mineral and Servant as this is, *Not* that I speak this to raise the Price of Coals on the poor, or others about *London*, or other parts where it is so much used, and especially in these times, when Taxes or other things may be more now, than at other times; No, for I fear they have not the Benefit of the lowness of Price as at the Pits; but I fear others have the Benefit of it, *who less deserve it*, and yet make the Poor and others, &c. pay dear enough for Coals, both at *London* and elsewhere, when, in Truth, I think (would they but lye their groundless pretences by) they may or might afford them cheaper to the Poor, &c. now than when they gave 11*s.* 6*d.* or 12*s. per* Chaldron at the Pits, or first Charge, besides Pittage, &c. but, Sir, I have observed something of the low Price of Coals for these Moderate Purposes, that the poor Miners who do dayly venture their Lives (as perhaps I shall hereafter have occasion to speak of more at large) may have Wages to live on, to get Bread for their Families by their Labour, and that those Gentlemen who have the good Fortune of such Coal-Mines in their Grounds, may be encouraged to set all Idle Hands to work, and prevent Increase of Beggars, and likewise may have a moderate Advantage by their Mines, whereas, I am affraid, those Owners who have their Mines up in the Country, and so have a long Carriage to the Water of their Coals, adventuring great Sums of Money in their Colleries, at the low Price Coals sell at now, cannot get much Profit, which indeed is not to be Comiserated by all well meaning Persons, for it is but reasonable where much is expended, that some Benefit should arise to the Adventurer, especially when such an Adventurer is so useful to the Publick, in imploying and maintaining so many Poor as this doth; nay, if Incouragement (or Profit) were not allow'd to the Adventurers, what would those Persons do to live so well that have Lands wherein Colleries are, yet have not a Fund to work such Colleries, being so obliged to let their Colleries to some Adventurers who have Cash? but if no Profit can be raised, I see no Reason why any Man should Adventure his Money. But to proceed, Lastly, there is another Bed of Coal, which is about New-Castle wrought yet not found on *Sunderland* River (as I know of) which is called the Stone-Coal, it is so called, because it has a sort of Stone, which is in the Bed or Vein of Coal; and about the middle of the Bed or Vein, yet it is accounted as good as the Main Coal, only it is subject to be a little Slaty, as I am told, and which I think cannot well be remedy'd, for it is hard to pick all Stones out, when wrought with the Coal, let Persons be never so careful, and now having given you an Account of the several sorts, names, and nature of these Northern Pit-Coals, I shall beg leave to give you a further account of Sinking these *Coal-Pits.* I am sensible you know when we sink a Pit, at first we break or cut the Ground four Square, and the Diameter of the Square is generally approv'd of to be, according to the Distick in *Virgil.*

Dic quibus in Terris, & eris mihi magnus Apollo,
Tres pateat Cœli spatium non amplius ulnas.

Virgil Eglog. 3.

About 9 Quarters and he hit it right, if he accounted an Ell according to the Flemish or Dutch Ell, which (I am told) is 3 Quarters, then we carry on our Works with such Timber as before mentioned, and in the same 4 Square Form, till we come near to the Stone-Head, where we form it into 8 Squares, the better to Incline it to answer the Ring, or round Form of the Stone-Work, which is always so wrought or fashion'd; but we always wish not to meet with any Quick-Sand or too great a Feeder of Water, whilst working through this Earthy Part to the Stone Head for some Quick-Sands have baulk'd our Winning the Pit, where we had begun so to Sinck, as likewise some great Feeders of Water have done, therefore many times we are forced (in Case of the Water) to sinck another Pit or Shaft to the stone, or lower shaft to have a Water-Course or Drift from the intended Coal-Shaft *Top Feeders best carried off, by another half Shaft.* to this other Shaft, that the Water which rises at the Coal-Shaft, may run into this Water or half Shaft, to be drawn there by Horses or Water Wheels, as there is a conveniency for it, that so the Water may not follow or descend, with Sinking; which Method I think is esteemed best to carry off Feeders in the Earthy, or top Part of a Coal-Shaft, because if the Feeders be of any considerable Quantity, it will melt, or Dissolve the Earth, so that it will drain through the Deals and Timber work, notwithstanding all that can be done; but *Quick-Sands* (if not to thick) are often *How to get* put through by Deals or Timber, for *through* we lay choak Deals (as we call them) *Quick-Sands.* which is Deales put in as fast, or all along, as we dig, the Sand, or Earth, that if possible we may keep the Sand by this extream quick Vigilancy and Care from getting *At Harraton,* in upon us, I have heard of Iron Frames *Durham.* that have been us'd (made square and deeper, then the thickness of the Quick-Sand) to put back these Quick-Sands, which may be of good use, tho' they must be dear to be so Cast or Wrought; but I do not find but this way of Timbering is most used, only they put stiff Clay to the Inside of the Timber work, which is forc'd and ram'd in next the Sand, to keep it back from the Timber.

And now am I come to the Stone-Head, or Stone Work, where we are likely to meet with hard Work enough, perhaps the Sincker, (or Labourer) has, as before, 12*d*. or 14*d*. a Day, yet it is not every Labourer who has been (by chance) in a Coal Pit, or at Labour in other sort of digging above Ground, that is fit to be imployed in this Work, but it should be one that understands the Nature of Stone and Styth, and Surfet, by some Experience had before in such sort of Labour, who for saving Moneys to the Owner, by imprudent managing, and breaking, and spoiling his Tools, or sinking Hacks (which we must have a Smith at Hand to be ready for sharpening, as the Points are broke off, which is still Chargeable and loss of time, or the Hacks wore off, with the constant Labour, can make no furtherance of Work, or does little; as also for his own Personal Security, for if he have not Judgment, the Shivers or Splints of the Whin or hard Stone, which is met with, will

An Expert Labourer to be employed.

Wound him severely (I have known it so) to the loss of their Eyes, and cutting their Leaders and Nerves, in Arms or Legs to their future Lameness and Inability of Labour ever after: But what is worse of all if he be altogether unacquainted with this sort of sinking Labour, he may loose his Life by Styth, which is a sort of bad foul Air, or Fume exhaling out of some Minerals, or partings of Stone (for there are several sorts of Stone, which we pass thro' as well as Minerals, &c.) and here an Ignorant Man is Cheated of his Life insencibly, as also he by his Ignorance may be burnt to Death by the Surfet, which is another dangerous sort of bad Air, but of a fiery Nature like Lightning, which blasts and tears all before it, if it takes hold of the Candle, which an experienced Labourer will discover and extinguish, tho' it be going to take at his Candle, and can sometimes smell to be Dangerous or hurtful, therefore all Sinkers should be skilled in these Matters for their own Security sake, as also for the Benefit of the Owner or Master of the Collery. for if £1000 or more be spent in carrying down a Pit or Shaft, almost to the Coal expected, and then by an Ignorant Man should be blasted by a strong Blast, by Surfeit; so that it may (as has been known) tear up your Timber Work and shatter the Gins, and shake the Stone Work, or Frame Work, so as to let in Feeders of

Water, besides the Destruction of the Persons in the Shaft, this would be a dismal Accident with a Witness, as well as loss of all the Labour and Cost by Ignorance.

Coal Owner.] But do you not meet with Springs or Feeders of Water in your sinking, which I reckon as great Enemys to the Prize and Concern, as those of Styth, Fire, and Whin?

Servant.] Yes, Sir, too often, and it is very rarely found, that a Pit of 40, 50, or 60 Fathom is sunk, without going through several sorts of Feeders; indeed, were it not for Water, a Collery in these Parts, might be termed a *Golden Mine* to Purpose, for *Dry Colleries* would save several Thousands *per Ann.* which is expended in drawing Water hereabouts, and I call them Feeders of Water rather then Springs, because, I have always observed, That the Water (which is the same in Taste, and as drinkable as what you call *Fine Springs* on the Surface of the Earth or above Ground, (which we so often cut through, and put back, or Cawke, like as a Ship or Vessel is Cawked on the Ocean or River) we by our Works find not to rise *upwards* or *Perpendicularly*, but to come in some parting of Stone, or Mineral, or otherwise, *between* or *parting* the Earth from the Stone, or else parting some kind of Earth from other sort of Earth, *slanting* or *sideways*;

so that we think it cannot be so well called a *Spring*, which doubtless some take to be, by Boiling always upwards, as if it came up through the whole Centre of the Earth; whereas (if rightly considered in our Opinion) it is impossible that Water should Perpendicularly so Bubble or Boil up, through such a vast Body of Rocks and Minerals, and therefore as we often go through and below those Waters or Feeders; it is plain it cannot come Perpendicularly, for we leave it above our Heads in the Shaft, either framed and cawkt back, or else we by a Drift or Water-Course, (as before mentioned) draw it at that depth and no lower, at another Shaft. Besides, it is further plain and apparent to us, that there are several Partings or Partitions, with several Concavities, where out comes sometimes Dry, Stinking, Sweet and Sulpherous Air, and such sort of Waters likewise, so that the Reason why Water Bubbles up, is, we think, because the Parting in which it had its Course is some how stop'd, and as it cannot descend through the mighty Body of Stone. it must of necessity rise through the spungy Earth, all which Water we suppose to come from the Sea, and so being fed by that inexhaustible Fountain, we call it by the Name of a *Feeder*, and that it may rise to the top of any Mountain or part of the Earth, we are subject to believe no great Matter of Wonder, because we are so often, by the Curious and Learned, told, That the Sea, this Fountain Head, is higher than the Earth ; and it is a common thing to see the like Performance, that any Water coming from the Fountain, will rise to the height of that Fountain Head, as in all Water Works, &c. in *London* and elsewhere.

Coal Owner.] Well, you tell me of Framing and Cawking back your Feeders and of drawing up your Water, let me hear your mannear of framing, and drawing your *Feeders* in your Shafts.

Servant.] For framing back our Shaft Feeders, we make use of Wood, but chiefly *Firr*, because it being a soft sort of, or spungy Wood, we think it swells with the Water lying against it, and Wedges best by reason of its yielding Quality or Softness, so that the least Thread, or Leak of Water, which may proceed from the least Chinks, or openess of the Wood, are best stop'd, and wedged back by Firr Timber, yet requires Judgment in the Head-Servant (or Over-Man) of the Sinkers, to give prudent Directions, how high or big to make this Frame, for by chance we may have a Feeder come in some small parting of Mettle or Stone, the Concavity whereof is not much more than half an Inch, yet having so much Liberty, will afford us a great

How to frame back Water.

deal of Water. Now if we were only to Wedge or Frame back that half Inch's Concavity, it is two to one but there may be a soft sort of Stone, or Mettle, above or below, (or both) that Cranny or half Inch Frame; which if, then will the Feeder certainly burst throw that soft Stone or Mettle, when as your Frame rightly order'd holds, or stands, so that your head Sinker, (whose sole Care above all things, must be to be always present at such Framing) should understand well what sort of Stone will hold back the Feeders with the Frame, and therefore before he puts in the Frame, he should know this aforesaid, and so cut out such soft Stone or Mettle, and making his Frame so much bigger at first and Judicially, may save a second Trouble and Charge thereof.

But it must be further Observed, That as this Mettle, or Stone, or Coal, may be ruff, and cannot be cut up so extraordinary smooth as to stop all Drainings of Water, then in such a Case we make use of Sheep-Skins with the Wool on, the Wool next the rough Mettle, &c. which being well Wedged in between the Frame and such rough Mettle, &c. the Wool we find to Perfect our Design of stopping the Water: And Wool is generally said to be of a very durable Quality, scarce ever to rot; I have known such Frames to have stop'd back whole

Waists of Water in Collieries, and not doubted but to last to Purpose.

Drawing of Water

And now as to Drawing of Water, we generally Draw it by Tubs or Buckets; whilst Sinking with Jack Rowl, or by Mens winding up the Rowl, or otherwise, if the Pit be Sunk more than thirty Fathom, then we use the Horse Engin, which Engine being wrought with one or two Horses at a time, as the Water requires, serves also, after we have Coaled the Pit, to draw up the Wrought Coals. Which Engin, tho' it be but of a plain Fashion, yet is found by Experience to be more servicable and expeditious, to draw both Water and Coal, than any other Engine we have seen in these Parts yet, notwithstanding we have had many pretenders, in many Kinds and Methods; though we would be glad any Ingenious Artist, could show us a better or more effectual way, for Expedition and Service, then we now use hereabouts. In some places we draw Water by Water, with Water-Wheels or long Axel-Trees, but there is not that Conveniency of Water every where, and as for Wind-Mills, or Ginns to go by Wind; 'tis sure the Wind blows not to purpose at all times, and therefore, whereas we cannot Sink Pits, but still will have some Water to draw, and many times so much as takes up all (or most of) the spare time we have from Coal-work; it follows we must have a Method whereby to draw this Water when we please, or at any time, or otherwise, we must continue this plain Method we now have, If it would be made Apparent, that as we have it noised Abroad, there is this and that Invention found out to draw out all great old Waists, or Drowned Collieries, of what depth soever; I dare assure such Artists, may have such Encouragement as would keep them their Coach and Six, for we cannot do it by our Engines, and there are several good Collieries which lye unwrought and drowned for want of such Noble Engines or Methods as are talk'd of or pretended to, yet there is one Invention of drawing Water by Fire, which we hear of, and perhaps doth to purpose in many Places and Circumstances, but in these Collieries here a way, I am affraid, there are not many dare Venture of it, because *Nature* doth generally afford us too much Sulpherous Matter, to bring more Fire within these our deep Bowels of the Earth, so that we Judge cool Inventions of Suction or Force, would be safest for this our Concern, if any such could be found that would do so much better, and with more Expedition than what is done generally here.

And now having given you an Account of our Method of Sinking Shafts, both through the Earth and Stone and Feeders, when I come to the Signs of approaching, or being near the Coal, we generally have the Roof, (or what lies next to, or upon the Coal) of a softer kind of Stone or Mettle, of a blackish Colour, yet do not desire, for the Good of you our Masters, Coal Owners, to have this Roof *too soft* or *tender*, neither to have a Feeder next the Coal or too near it, in a too tender Roof, for Water spoils Coals in the Colliery for Sale, making them black and heavy, which doth not relish well with the Ship-Master or first Buyer, because he buyes here by Weight, and therefore being heavy, a less quantity of those Wet and so heavy Coals loads him, then of dry bright Coals, which he finds in his making out in Sale at *London* or elsewhere, where he sells by Measure, and not by Weight.

And again, if the Roof, or Cover, of the Coal be too tender or soft, though dry, we must not take so much of the top of the Coale away, as when the Roof is strong, but leave perhaps about a Foot thick of the Coal top for a Roof, least by the softness of the Mettle Roof, that Roof should fall down and kill your Miners, or what is also bad, bring a Thrust, or a general Crush in one of your Collieries to close it quite up, and thereby lose the Colliery; it is there fore of great Advantage to have a stong Roof or Cover of the Coal, because of what is already said, and further, because (though it is but rarely so) you may take away, and use all the Coal to the Roof, which is a great Increase of your Quantities, wrought out of such Pits: but now being happily past all this Trouble and Charge, after the prescribed Methods, and having happily Coaled this Noble Main-Coal. My Business of a Sinker is at an end, only I must give you an Account of the Thickness of the Bed, or Vein of this Coal, which (as before) may be about eight, or eight Quarters and a half thick, and then lastly making a Sump, as before mentioned, I must take my leave of this Subject of Sinking, after you have been pleas'd to give your Sinkers (because it is customary) the Labourers, whom I have employed for you, a Piece or Guinea, to Drink the good Success of the Colliery, which is called their Coaling-Money, and then lye Idle till you have occasion to break, or begin to Sink another Pit, which I hope will not be long first, for it is judged to be a Point of Wisdom and Care, not be too long or tedious in providing a Pit or Pits, to be ready sunk and ready to set to Work against the time you have Wrought out your Coaled Working Pit; that so you lose no Time, or Charges of the Water-drawing, by Coal Work, standing some time before you Coal another Pit. And now I must leave your to your Viewer, or Head Under-over Man, who is to take Charge of a Regular Working of the Colliery.

Coal Owner.]   Well, honest Sinker, now thou hast Coaled the Pit, take thy Wages and the Labourers too, with what thou sayest is Customary to drink, and this Viewer, or this under over Man thou hast Recommended to me, I hope will give me an Account of what his Business must be, and how he designs to go on with his Coal-Work, being, I perceive, you Sinckers differ in Judgment, and Methods from hewing or working Coals.

Viewer.]   Sir, my Neighbour and Acquaintance having, in my Opinion, provided a very good dry and convenient Shaft; with your good Leave, I shall make bold to acquaint you of what is requisite to be done on yours and my part, for there is something on both sides to be done or provided: In the first Place, Sir, on your side, you are to buy in a stock of able strong Horses to draw your Coals to Bank (or Day) out of the Pit; and whereas by *Of Lineing.*   my Plumet-Line, which I Measure the depth of the Shaft by, as they line or sound for the depth of a River, I find the Pit is about 60 Fathom, about 21 Scores of Corves 21 *Scores a*   of Coals wrought, and drawn *per* Day, *Days Work.*   is a good Days work, and as much as most Collieries of that depth, can, or do constantly Work, which Pit will require at least eight Horses every Day, to perform that Work, which is, as always Customary, 4 shifts of Horses, at two in at a time, and indeed you shou'd have a spare Shift, or two Horses more ready, least a Horse or two, Tyfle, or be out of Order by a Fall, or any other Accident or Illness, and so out of Condition to Work; for if you want a Shift or two of those eight Horses, you either loose a whole Days Work of Coals, or at least one $\frac{1}{4}$ of it; besides, your Work-men, or at least some of them, will expect as much for their Days Work, though it wan't a quarter of its Quantity; so that two Spare Horses, which so you know, makes it up ten Horses, is, in my Opinion, for those Reasons above, rather more Convenient for you to buy for one Pit, and it will Ballance the Charge of maintaining a spare Shift of Horses, in the said two Respects, both of supplying a lamed or sick Horses place, (which would be destroyed by work, when in such an ill and unfit Condition, as also of making out your Days Work, and wages as afore-said. These ten Horses must at least in these Parts stand you in six or seven Pounds a piece, or more the better; I have known some do not value to give ten or twelve Guineas a piece for young, strong and mettle Geldings or Mares; for by those Qualifica-tions you may expect them to last longer in this heavy Work, and old weak Horses (though cheap bought) are soon wrought out or spoyl'd by such

Labour ; and we do not find Stone-Horses fit for this use, because they are more unruly and ungovernable, neither can we in these open Countries conveniently turn them out in Summer time to Grass amongst themselves, or Geldings, least they lame or spoil one another, and it is very convenient to refresh these constant hard working Horses in their Limbs, in Summer time, otharwise they are subject to Beat or Founder to their Feet or Leggs ; *Two Sledge-Horses.* we must have two more Horses of a less Value, bought to Sledge out with, or draw the Corves as they come out of the Pit, on a Sledge on both sides the Pit, of three or four Pounds a Piece, or Aged wore out Coal-Horses, which after some time Wrought, you will have, may serve turn for Sledge Horses ; the same Gin that your Sinkers used, will serve for Coal Work, and every thing thereto belonging : But a Cable of three Inches round and of good Stuff, will do better for Coal-Work, because it will be lighter for the Horses, than this Sinking Rope or Cable of four Inches, or thereabouts.

And whereas I speak of Corves, or Baskets to put the Coals in, we must have a Man (which is called the Corver) to make them.

The Corves to be made of Hazle.

He must have a good Quantity of young Hazle Rods, provided for that Purpose, with young Plants, or Sippleings, as we here call them, of Oak, Ash or Aller, of about three Inches thick, or
better, for the Corf-Bow; we buy the   *Corf-Bow.* Rods by Bunch each Bunch, contain-
ing about a Hundred Rods, at about Six-pence *per* Bunch, and the Bows being better than two Yards long, for half a Crown or three Shillings *per* Dozen, or thereabouts.

Your Corver ought to be just to you, in keeping up your Corfe, for with Working the Coals, being drawn pretty briskly up, the Corves are subject to Clash and beat against the Shaft sides, and so beats down your Corfe dayly, that if your Corves be not dayly beat up and mended ; you may lose more than one Inch dayly, which would bring your Measure or Corfe, of fourteen or fifteen Pecks, down to nine or ten Pecks, and so lose you a third of your Measure, and cost of Working or Hewing, *&c.* for it is most usual to agree with your *Hewers* of Coals or Miners, by the Score of Corves, by chance for ten Pence or twelve Pence for each Score, according to the tenderness or hardness of the Coal, or according to what the Mine will afford, and not by the

Day, or *Shift* Work, for it is common to give about twelve Pence or fourteen Pence for each Shift, when perhaps you will not have above thirteen or fifteen Corves a Man *per* Shift; so that is clearly best to agree by the Score, and then good Hand, good Hire, as we say, and you pay for no more then you have Wrought, or comes out of the Pit.

It is the Over-Man's Business to place the Miners in their Workings, or Boards, or Headways; And every Hewer gives him an Account of what Quantity he will Work dayly, or give; so that he every Day taking the Account of the Work, finds what Number of Men will Compleat the intended Day's Work or Quantity.

Besides these Miners, called Hewers, there is another sort of Labourers which are called Barrow-Men, or Coal-Putters, these Persons take the hewed Coals from the Hewers, as they work them, or as fast as they can, and filling the Corves with these Wrought Coals, put or pull away the full Curves of Coals, which are set, when empty, upon a Sledge of Wood, and so halled all along the Barrow-way to the Pit-Shaft by two or three Persons, one before and the other behind the Corfe, where they hook it by the Corf-Bow to the Cable, which, with the Horses is drawn up to the top, *or to Day*, as it is their Phrase, where the Banck's-Man, or he that guides the

Sledge-Horse, has an empty Sledge to set the Loaden Corfe on, as he takes it out of the Hook on the Pit-Rope, and then immediately hooking on an empty Corfe, he leads his Stead-Horse away with the Loaden Corfe, to what Place of the Coal heap he pleases.

One of these two Men that guides the Sledge-Horses, on the *Banck* or Surface of the Earth, is called the Over-Man of the Tree, or chief Banck's-Man, for he takes an Account of the Quantity of loaden Corves of Coals, which come to the Bank or out of the Pit every Day, by Sticks or Pieces of Wood, which is put in by the last Barrow-Man of the Number which is under Ground, so that if his Memory fails him, those Sticks immediately show him how many Rounds the Barrow Men have put, and so what quantity of Scores are Wrought: And if there chance to be any Hewers, who do not Hew their full quantity, they give first an Account of it to the Under-Over-Man, as before mentioned.

Then the Work falling short of the set Day's Work of twenty or twenty one Scores, that Abatement is taken Notice of by the Banck's-Man, and is made good another time, or else at the general Pay, (being noted by the Clarks of the Works, to whom the Banck's-Man dayly gives an Account of the quantity Wrought) it is abated on these Persons that made the Abatement, that so the Coal-Owner may not pay for more Coals than be had.

And further, there is strict Notice taken dayly by the said Bancks-Men, if Honest, of the filling of the Corves with Coals, for otherwise both the Hewers, and Barrow-Men, will confederate under Ground, and if the Coals be Hewed or Wrought pretty Round and Large Coals, they will be sometimes so Roguish, as to set those big Coals so *Fraud.* hollow at the Corfe bottom, and cover them with some small Coals at the top of the Corves, and make it look like a full Corfe, which Fraud when discover'd by the Banck's-Man, and the Coals shaken in close appears to be little more than a half Corfe, is Noted by the said Bancks-Men, who do not empty that false Corfe, but setting it by as it is, when shaken in, lets it stand on the Coal-heap, till the Offender comes to Banck or from under Ground, and there is Reprimanded Publickly before others, and the forfeiture is to give another full Corfe, with that bad filled one for it, or otherwise at the general Pay the Offender Forfeits six Pence *per* Corfe, for every such bad filled Corfe to the Owner, which is Deducted out of his Wages by the Steward or Pay-Master.

So that the Over-Man of the Tree, or Chief Banck's-Man, has two Pence for each Day more then the other Bancks-Man, for taking Care of the Quantity Wrought, and of the bad filling of the Corves, the Common Wages is about sixteen Pence a Day for the Over-Man of the Tree, aud fourteen Pence a Day for the other Banck's-Man, or Sledder.

The Wages for the Barrow-Men is usually about twenty Pence, or two and twenty Pence a Day for each Tram (that is to say) for putting so many loaden Corves, as are carried on one Sledge, or Tram in one Day to the Pit Shaft, at the first beginning of Coal Work in a new Pit, it is usual to begin with six Trams, which put and hook to the Cable three Score and ten Corves a Piece every Day, which makes up twenty one Scores a Day, and then about every thirtieth Day, or so, afterwards, they have a Barrow-Man, or Tram added, still at the same Wages of twenty Pence, or two and twenty Pence *per* Day, till by chance the Coal will be so far Wrought under Ground, that there must be sixteen Trams; which brings it so, from seventy Corves put, to about twenty six Corves a Tram, so that the more and further a Pit is Wrought, you see the dearer she lies in the Charge of Barrow-Men, or putting, for they still keep up the first Price of a Day's Work, were you to add never so many Trams; nay sometimes a Pit may happen to have a Hitch or Dipping of the Thill or Bottom of the way, by which such Setling or Descent, you are sensible to pull a full Corfe, up such a sort of a rise requires more Strength, then where it is drawn down Banck, why then in such a Case these Barrow-Men will rise their Wages two Pence a Day for each Tram, or more according as the Hitch is, therefore it is always look'd upon to be of good Advantage to the Colliery, to have a rise in the Thill, and of the Coal, as we work, because it is easier in putting, and so consequently Cheaper to the Owner.

Why the Rise is.

Here I cannot but take Notice by the by, of what we generally have in all Collieries, and that is, three is a Rise, or Ascent, for a Colliery under Ground, and so by Consequence the contrary way a Dip or Setling for a Current; this said Rise is commonly found to be to the South, and to the South West of most Collieries. We can give no Reason why this should be so, only I have with my self Fancied it may be (by a Magnetic sort of Quality) with Respect to the Sun's Noted Ascent to the said South Points, and I call it Magnetic, not in Respect of the Magnet or Load-Stone, because that draws to North Point but Magnetic, *quod magnas habet Vires.* Or otherwise by bearing some Proportion with the Nature of the Descent and Current of the Waters, or Rivers in this Kingdom, which are generally Observed to be to the Eastwards : but I shall wave this piece of Curiosity, it being above my Capacity, and shall proceed to give some Account of what is a Viewers Office and Duty, and how he ought to be qualified for such an Imploy.

And First, this sort of Servant (as in all other Cases) ought to be well skill'd in this great Concern he takes in his Hand, he ought to know *Lineing*, and *Levelling* well, as also the Method of *Coal-Working*, together with the Knowledge of the *Nature of the Coal ;* for there is very great Occasion for all these four Qualifications, as follows in the Sequel, he should know how to Line well, because he must set out his first Work or Headways, according to Rule by the Compass, (which is as useful under Ground to know which way and how far he goes, as it is in Sailing) nay, being more in the dark (because so deep in the Earth) it is impossible to know how we go under Ground, or at least so exactly as requisite, unless we have our Compass, and so Line by it, to any Point thereof, which we by the Scituation of our Ground and Colliery are obliged to Work to, so that by the Compass seeing how the Headway lies (which alters pretty much in some Collieries perhaps more to the South-West or South East, for the South Headways) then in others, this Headways, I say, or first working (which is generally Wrought by the Yard, whether it be Single or Double) is carried on, according to the Grain of the Coal, as it lies along the Grain, and not cross the Grain, neither is it Wrought so wide, as the other Works or Boards are; a Yard and Quarter broad or wide for a Headways is full sufficient, and out of this it is, we turn off the Boards or other Workings, for every particular Hewer, or Miner, and that Board or Work-place for that one Man, is generally about three Yards, or better, in breadth, and so from the Headways, which we will say runs South, we work the Boards East and West of the Headways, not taking all the Coal away as we go, for then we would bring a Thrust of the Collery upon us: But after we have carried our Head-ways Drift, about eight or ten Yards from the Pit Shaft, then we consider of a Winning, how much to allow for a Winning, which is about seven Yards in these parts, or otherwise, according to the Quality, or Tenderness of it, more or less, as by Judgment is thought safest and best; out of this Winning of 7 Yards, perhaps we dare not venture to take above three Yards breadth of Coal for a Board; so that then there is but three Yards for one Man to work by himself, and therefore would be Dangerous for two Persons to Work together, least they should strike their Coal-Pics into one another, or at least hinder one another;

then the Remainder of four Yards is left for a Pillar to support the Roof and Weight of the Earth above ; which makes it out so, that there is not quite half of the Coals taken out of the Ground which lies there.

And in this Method it is that we carry on this Coal Work, taking away three Yards or better, according to the Strength or Softness of the Coal, and leaving four Yards standing for Pillars to support, as before, so that reckoning seven Yards to a Winning, and to have twenty Boards, we must of Course drive our South Headways 150 Yards in Length, or thereabouts, to have all the Boards of one side of the Headways: And therefore in a working Colliery, it is best to win out the Headways first, so far, as that we can clap a whole Set of Hewers in to Wine for the Boards, at first entring into a New Pit, for then we keep the Work going all on one side, and when that side is Wrought out, we begin *de Novo* to work on the other side, the same Headways serving us again as before, and besides we bring our Barrow-Men to put three Score and ten Corves a Day *per* Tram, as we did at first on the other side of the Headways ; which is to the Owners Advantage ; and after this South Headways is thus Wrought out to the East and West Hand of it ; then it follows to turn the Workings to the North Headways, or other side of the Shaft, which we go on with after the same Method, as in the South, till we have wrought by these same Rules, all the Coal we can, with safety venture to Work or take away ; and so by chance have Wrought, or waisted the Colliery 8 Score or 200 Yards to the East, West, North, and South of the Pit-Shaft, and then it is time to have another Shaft at that Distance sunk for another New Pit, which if happily done by the dayly Care, Prudence, and Orders of the Viewer, and his Drift to the New Pit, carried on exactly so as to hit the New Shaft, and supply her with Air, Then has he evidenc'd both his Care and Parts in the Respects I have mention'd, and well deserved his 15*s.* or 20*s. per* Week, or more, as he has Pits to look after ; and he has an Under-Over-Man, always allowed for every Pit, whilst one Viewer serves for three or four Working Pits, and the Over-Man's wages is about 8*s.* a Week.

But now I would give you an Account, that for the fafety of the Miners, we must be careful of guiding the Air under Ground, least we bring a Styth, or a Blast, by the Sulpher or Surfeit, upon the poor Men, as of late it did, not far from *New-Castle*, I think it was but in *October* 1705, that I was told by one who was acquainted with, and see some of the Dead buried, and had been at the Pits after the Blast, that

About the 3rd or 4th Day.

there was above thirty Persons Young and Old slain by a Blast, perhaps in less than a Minutes time. How it came to pass he could not give me an exact Account, any further then by what the Banck's-Men, and those who were about the Pit, and heard the blow, and see what it threw out of the Pit, and shatter'd about the Gins: There was one thing very Strange in it, as I was told, That a Youth of 15 or 16 Years of Age, was blown up out of the Pit and Shaft, and carried by the Blast about 40 Yards from the Shaft, the Corps was found all intire, save the back part of his Head, which was cut off, though the Shaft is said to be odd of sixty Fathom deep, which is an Argument of the mighty Force this Blast is of; but those Over-Men or such, who should have given an Account where and how it first took, were all Slain; yet this is known by woful Experience, and which I myself have seen and narrowly (by good Providence,) escaped that some Collieries are very subject to this fatal Surfeit, and therefore it behoves the Viewers and Over-Men to be experienc'd in guiding the Air to good Purpose, as also to Order well and Prudently for Styth, which I before spoke of, doth Destroy the Ignorant and Unwary.

And thus it is plain, that both the Officers and poor Miners, are in dayly Peril and Hazard of their Lives, for a poor Livelyhood, and that they may be easily Destroyed by Ignorant and Unskilful Managers, from which sudden and sad Misfortunes, I heartily Pray, *Libera nos Domine:* I have (as before spoke of) observed that their Workings or Boards are carryed by single Persons, a pretty long way under Ground, yet they frequently hole, or cut through from one Board to another, to carry their Air forwards with their Works, and to the end or Face of their Boards, or otherwise, the Air and Communication of it would not be good and safe to Work, but this Care of the Air must be taken in general, That it be not too much Dispersed, or too much Liberty given for want of Stoppings to spread it self from the particular Workings we are in Hand with, for the brisker it ranges in the Works, the sweeter and safer it is for the Miners.

Coal-Owner.] But, Friend, can you give me now any good Account of the Nature of the Vend of Coals, as well as of the Working them.

Servant.] Yes, Sir, I had some Experience in that, not long since; and shall therefore give you the best information I can of it, especially on the River *Were* or *Sunderland* River, where I observed. That, first, you must have a good Stock of Coals provided against the time of Sale, which is chiefly in Summer by Reason of the Weather, which makes it hazardous for Ships to Sail in Winter on those Coasts, for which Reason, I have heard good Saylers say, they had rather run the Hazard of an *East-India* Voyage, then be obliged to sail all the Winter between *London* and *New-Castle*, whither it is not common for the Men of War to be out Convoy all the Winter, on those Coasts, perhaps for the same Reasons.

And though for such a Stock of Coals, it be at first considerable to Disburse, if it be but two or three Thousand Pounds, yet it will afford you good Interest for it, in two Respects, first you get your Coals cheaper Wrought in Winter time, then in Summer, or time of Trade, because Labourers or Miners are then more Numerous thereabouts, being Land-Sale Collieries are most commonly laid Idle for that Season of Ill Weather, and rather than those Labourers will lye Idle, they are prevailed on generally, and as it is customary to lower their Wages during the Winter Season; and then, Secondly, by having a good Stock aforehand, you have wherewith to Answer Demand, in time of Vend, so that thereby your Return is greater and doubtless (according to the old Maxim) the greater your Return is, the greater will your Profit be; for how, if I have a Shop, and have no Goods in it, shall I furnish the Demand of my Customers, when they come to buy of me, they will say they cannot stay till I get them bought in, or made: and surely

Ships can as ill lose time, or tarry till you get a Stock of Coals Wrought, because it is at no time Prudence, or proper to lose any part of the trading Season, but surely it is much more imprudent to lose time, when for safety the Ships dare not Trade without Convoy in this War time: So that to lose time then, is to lose the Opportunity of Convoy, and by consequence to loose Trade, which I know to be too true by woful Experience in my late Business, and therefore, I hope you will Pardon this Advice from your Servant, being he does it to forwarn you of the Mishief and Damage, of want of a good Stock laid in aforehand, and I hope you will not be so unreasonable as to think, Brick can be made without Straw, or blame us, as others have done, because you will not provide for, or get a Trade, when you might thus secure it, and certainly it is the Master, and not the Servant, that should advance Moneys for a Stock or Trade, which can never be had, without you make such Provision for it, though it be the best Colliery in the whole World; but to add this is needless to any Gentleman of common Sence; therefore I shall proceed to my next, that an Interest is to be made in the Fitters, or those Persons who live at the Ports and have Keels, (which are much like to Lighters Built) to load the Ships, for the Rivers are not Navigable for Ships, so high as they Keys or Coal-Steaths, therefore these Keels are built flat Bottom'd for the purpose, and carry the Coals on Board; these said Fitters you are, I say, to have an Interest in, for they are best acquainted with the Ship-Masters, and many times, nay generally have parts in the Shiping, so that to be sure where they Adventure their Moneys in a Ship, there they, in Reason, expect to have the Benefit of Loading thcm as oft as they can, and so it is that Fitters have grcat Interest in loading the Vessels, and by consequence can befriend what Coal-Owners they please in the Vend of Coals; and of late I fear it is, that they knowing their Interest and Power, make some Advantage (especially of new Collieries) to bring Trade to beginners; their pretence is to have and get no more then two Shillings and six Pence *per* Chaldron of the Master for Fittidge, which because a Keel carries no more than seven Chaldron a piece, at *Sunderland*, is but seventeen Shillings and six Pence *per* Keel,

for the rest of what they have of the *Keel-Men*. Master of the Vessel, is given to the

Labourers that work the Keel, but (as before) for the Encouragement of Trade, I am afraid you must abate six Pence *per* Chaldron of the pretended current Price privately, either by doing as it is said, some Persons have formerly done, that is, by ordering your Steward to return the six Pence *per* Chaldron, when the Fitter pays him, and so may say, as they did, you don't do it, because the Servant did it by private Orders, which the kind Fitter will keep so close and secret, that he would not Devulge it, though he give broad Signs of it to another Owner, when he complains of want of Trade with him, but yet for fear it should Create open War amongst the Owners, he for his own Interest, will not publickly declare he has such an Abatement from such an such a kind Owner, of which Number, if you can by such a Method get to be one, than may you reasonably expect a Vend, and the Fitters Favour, because it is his Interest.

From hence it is that one great Colliery on the *Wear*, is thought to get so much of the Fitters Favour, and (Trade, much good may it do at the *Current Noble Price* Coals go at there now;) but I have heard some say, which seems more reasonable, they would desire but a reasonable Living-Price for Coals, and have all to be Just to such an Established Current Price, that so the Owners of Vessels, may not run away with all the Profit, whilst the Consumers, and especially the Poor and the Coal-Owners get nothing by it; and this puts me now in mind of an Act of Parliament which was in Force, and Observed, for Admeasuring those Keels spoken of already, which I am told was design'd for a just Gauge or Measure, which should Oblige all Persons to be Content with, but whether the said Act did Oblige Owners not to give any more, I cannot hear yet.

However it is a great Complaint that there are several Keels which work on that River *Wear* (if not on *Tyne*) which are not Admeasured according to that Act, and that the Persons that did formerly look so well, after such Admeasurement, have not of late done it as ought to be, and all is for want of such Admeasurement as the Act required, but indeed I have never heard of any Complaint of this, either by Master or Owner of Vessels, I suppose it is because neither of them care how much Measure they get, nor how low the Price of Coals be there, so they be but dear enough at *London*, or where they are sold again.

But, Sir, in time you will best Judge of this Matter, after you see what Profit your Colliery gives you, I would think some Owners have already had more Reason then myself to speak and stir more in it, for the Motive I have is my desire my Masters should live, which would be Encouragement to them to pay us Servants more Chearfully, for I have found that where Profit doth not arise there Wages are paid, (though we venture our Lives never so much) but very Grudgingly, if at all.

But there is one thing more I would willingly take Notice of before my Conclusion, and that is the ill Condition of the Harbour of *Sunderland*, it is Confess'd that they have as good, or better Coals on that River, then on *Tyne*, and so doubtless might have as good a Vend, or better then *New-Castle*, because it is nearer to *London* than *New-Castle*, and because of the Commodity. Nay, farther, because there is better Measure allowed they say, for at *Sunderland* they pay but for seven Chaldrons, when they have actually eight Chaldrons, whereas at *New-Castle* I have heard them say, if eight be on Board the Keel, they are paid for that eight, and so do not give eight for seven, as at *Sunderland*; yet for all this, *New-Castle* gets the Majority of Vend by much, for if any Storm arises at Sea, there is no safety in offering to go into *Sunderland*, there wants a Peor, as at *Whitby* and *Burlington*, or elsewhere, besides the Bar is so choaked up, that there is great want of Water; so that were there something built for Security of Shipping, and cleansing the Haven there, it would bring doubtless much more Trade to it, then it now has, and there is no fear but were this Recommended to the Consideration of our Honourable Redressors, they would take effectual Care of it, for thereby both Ships and Lives would be saved, which are too often lost, for want of such Safe-Guard, and indeed, why should not the Coal-Owners and Fitters (who are Interested by Coal-Trade) be obliged to contribute towards so useful and necessary a Work as this is; and surely one need not fear but the Masters of Vessels will be as ready for their Lives and Interest Sake, as any body, and be likewise as free to pay for there Safety there, as in *Burlington*, or any place else.

I have heard that the Annual Vend of Coals at *Sunderland*, is Computed at about sixty Thousand Chaldrons, if the said but sixty Thousand Chaldrons were to pay four Pence *per* Chaldron, it would raise a Thousand Pound *per Annum*, whereas it may by such amendment and safety of the Harbour nigh double the Trade, yet this Thousand only *per Annum* would do the Business I am told, as also with what other Additions and Allowances of Tunnige for other Wares and Merchandize as are paid at the Ports aforesaid.

The said four Pence *per* Chaldron might be raised thus, and I am apt to think nobody would contradict it, *viz.* the Coal-Owners to pay two Pence *per* Chaldron, the Fitters one Penny *per* Chaldron, and the Masters of Vessels loading in that Port, one Penny *per* Chaldron, which doth the Work.

Object. Ay, but what will *New-Castle* say to this? we know they will say it will prejudice their Trade, and therefore they must be against it with Might and Main.

Answ. Why truly, I see no Reason, but one Part of the Kingdom may and ought to be encouraged as much as another, and that the Adventurers and Poor thereabouts should expect Trade and a Lively-hood, as well as at *New-Castle* ; and besides, this is but for one Branch of Business, namely, Coals ; but surely again if it be the Preservation of so much Shiping, and sefety of many Lives. No Christian Man would be against such a good Act, for then we should not hear of the Complaints of so many poor Widows and poor Fatherless as we do thereabouts ; and what Wrecks have been seen near *Sunderland,* or about a place called *Souter* between *Sunderland* and *New-Castle,* those Northern Parts can Witness, whereas had there been Provision of Safety made in that Port, those Misfortunes or Losses had been prevented.

The Conclusion.

Thus, Sir, have I run over this small Piece with Brevity, and as plainly as I could, in hopes some other able Pen will give it some Lustre, and to the best Advantage, it being not of the least Moment ; and now I must remind you of the Custom of these Miners, that as soon as the Coal-Pits are Coaled, and Coal-Work begun, these Miners, &c. expect something to Drink, which is sometimes 5 or 10 Guineas or more according to the Generosity of the Owner.

Pecuniae Obediunt Omnia.

F I N I S .

THE ROMAN WALL

All these books are lavishly illustrated

HADRIAN'S WALL. Guide to the Central Sector by *Robin Birley*. Forty-eight pages. **60p**

THE ROMANS IN NORTH BRITAIN by *T. H. Rowland*. Sixty-four pages. **80p**

THE ARMY OF HADRIAN'S WALL by *Brian Dobson* and *David Breeze*. Forty-eight pages. **60p**

BUILDING THE ROMAN WALL by *Brian Dobson* and *David Breeze*. Revised and with new illustrations. **60p**

CIVILIANS ON THE ROMAN WALL by *Robin Birley*. Sixty-four pages. Illustrated. **£1**

SHORT GUIDE TO THE ROMAN WALL. Fifty-six pages. A detailed and up to date guide with numerous maps. Remarkable value. By *T. H. Rowland*. Illustrated in colour. **60p**

HOUSESTEADS ROMAN FORT by *R. Birley*. Four pages in colour. **35p**

CHARTS OF ROMAN WALL. A series of magnificent coloured reconstructions of the Roman Wall. Size seventeen inches by twenty-five inches. Price **80p** each.
1. **South Gate, Housesteads.**
2. **Mansio or Inn at Vindolanda.**
3. **Latrines at Housesteads.**

VINDOLANDA IN COLOUR. A magnificent souvenir of the Roman fort and settlement which has attracted so much attention recently. **£1**

THE ROMAN WALL RECONSTRUCTED. Twelve magnificent paintings in full colour showing the Roman Wall as it was when built. Paintings by *R. Embleton*. Text by *C. M. Daniels*. **£1**

DERE STREET. Roman Road North by *T. H. Rowland*. From York to Scotland. **70p**

HOUSESTEADS IN THE DAYS OF THE ROMANS by *R. Embleton*. Colour. **60p**

WHAT THE SOLDIERS WORE ON HADRIAN'S WALL by *H. Russell Robinson*. colour. **£1**

BIRDOSWALD FORT. by *Peter Howard*. **60p**

VINDOLANDA JEWELLERY. by *Martin Henig*. **60p**

VINDOLANDA WRITING TABLETS. by *A. Bowman* and *J. Thomas*. **60p**

CHESTERS AND CARRAWBURGH in the days of the Romans. **70p.**